Microbiology, immunity and ant

By Pip Flowe___

Contents

Introduction

This book is an updated amalgamation of two earlier books in the Basic Introductions series called *Basic Introductions - Disease and immunity* and *Basic Introductions to Biology - Bacteria Viruses and Fungi.*

This version gives some essential background about bacteria, viruses and fungi, with particular emphasis on their relationship to humans. Reading on, you learn about the body's immune system and how it deals with pathogenic microorganisms (including viruses). After a thorough study of the system you then discover something of the different ways that antibiotic, antiviral and anti-fungal medications work. The book does assume some basic knowledge of cell biology as described in Book 1 of the series. However there is a glossary at the end to help when unfamiliar terms crop up.

Bacteria

"Life on earth is such a good story you cannot afford to miss the beginning... Beneath our superficial differences we are all of us walking communities of bacteria." (Margulis & Sagan, 1986)

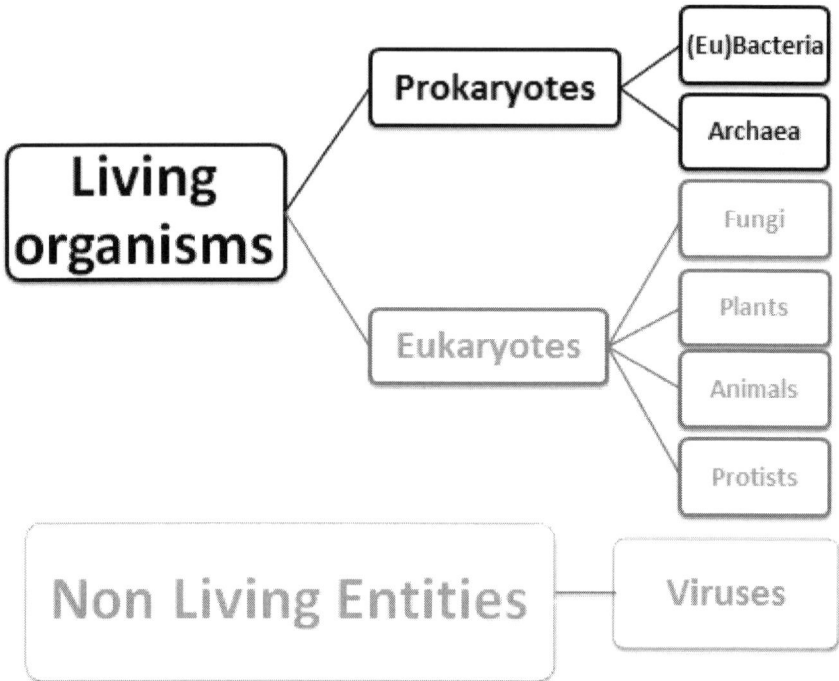

Where are they found?

Bacteria have shown themselves capable of flourishing in any environment on earth. There are bacteria deep in the earth's crust, on the highest mountains, in the ocean deeps and inside and on other living organisms. It is not surprising they are such tough organisms when you understand that the primitive bacteria called Archaea appeared just a short while after the earth started to form. Conditions then were hostile – a methane rich atmosphere(no oxygen) and unscreened UV light from the sun. The living archaea have earned the name 'extremophile' due to their ability to flourish in such extreme environments as hydrothermal plumes where the temperature approaches 500°C. Other Archaea can tolerate toxic chemicals and gases.

Bacterial species are usually found in hypertonic environments – their strong cell wall helps resist bursting. Some species (e.g. mycoplasmas) however thrive in isotonic conditions.

All bacteria can carry out the seven characteristics of life; respiration, feeding, excretion, reproduction, movement, sensitivity, growth and are therefore classed as living.

Bacterial structure
Bacteria(and archaea) are classed as prokaryotes. At one time the two groups were lumped together, but research has shown significant biochemical differences between the (Eu)bacteria and the Archaea. Accordingly there are now two groups of prokaryotes as shown on the diagram. Members of both groups mostly range in size between 1-10µm. The exception is a group of tiny bacteria called mycoplasmas, measuring between 0.2-0.3µm. Note there are one thousand µm (micrometres) in a millimetre. Unlike eukaryotes (plants,animals,fungi etc-see above) prokaryotes lack a nucleus, meaning their DNA is naked in the cytoplasm.

Bacteria also lack membrane bounded organelles such as mitochondria, ER, Golgi etc. meaning they lack the compartmentation present in eukaryotes. It has been discovered that the loop of DNA is somehow confined to a limited area in the bacterial cytoplasm. The lack of compartmentation does not make bacteria less efficient – they carry out functions differently, in some cases using specialised areas on the inner face of the cell membrane. The diagram summarises some of the features of bacterial cells in general.

Some species have a second membrane outside the cell wall (but never the Archaea)

All bacteria have a cell wall surrounding the cell membrane

No nucleus-DNA in a loop or plasmid

Some bacteria have flagellae for swimming

No membrane bounded organelles are present

All have free ribosomes

A gelatinous capsule may be present in some species

All have areas of membrane specialised for various functions

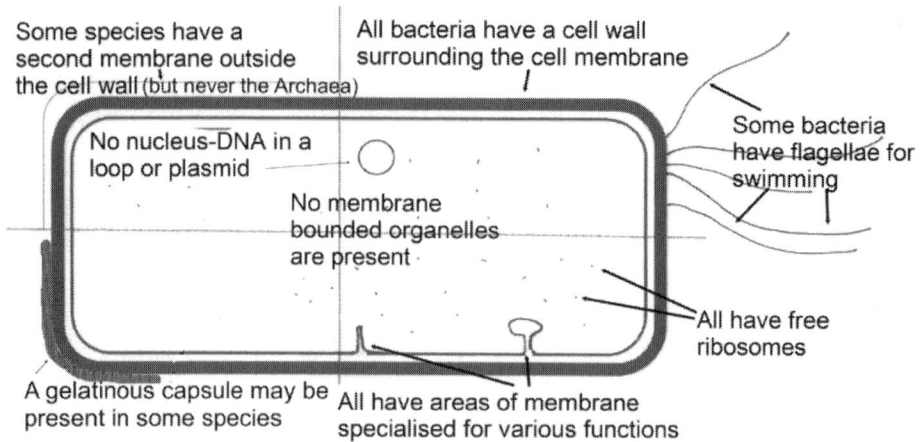

Here are some more key points.

Cell walls –all species of bacteria except mycoplasmas have a cell membrane surrounded by a cell wall. Some species of (eu)bacteria have a further membrane outside the cell wall. (See heading for Gram+ and Gram – later). Archaea never have this outer membrane. Unlike plant or fungi cell walls (containing cellulose or chitin respectively), (eu)bacterial cell walls comprise a substance called peptidoglycan. Peptidoglycan is a complex mesh of sugars and amino acids. However the archaea do not have peptidoglycan as the basis of their cell wall – some have a protein rich wall, while others have a higher proportion of sugars.

Cell walls of bacteria fall into two types, called Gram positive and Gram negative after the differential staining method used to aid identification.

.... Outer lipid membrane
Cell wall
Inner lipid membrane

Cell wall
outside
cell membrane

GRAM POSITIVE
BACTERIUM

Cell membrane
outside
thinner cell wall
outside inner cell
membrane

GRAM NEGATIVE
BACTERIUM

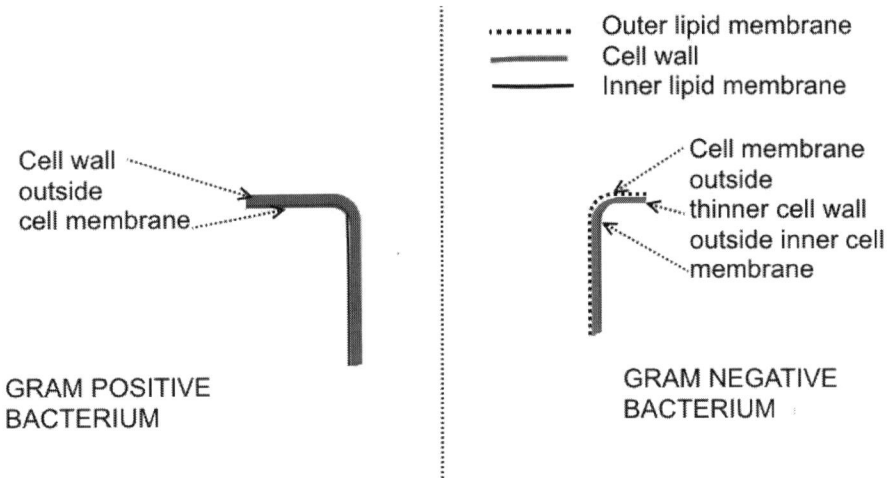

You can see that the Gram Negative species have three layers compared to two in the Gram Positive.

Ribosomes – all prokaryotes have ribosomes. These are smaller than the ribosomes in eukaryotes but do the same job. Interestingly it has been established that the fine chemical structure of archaean ribosomes is closer to that of eukaryotes than of the (eu)bacteria. (Londi, 2010)

Flagellae and pili – Where present, flagellae are the swimming 'tails' of bacterial cells. They lash around driven by protein motors in the cell membrane. The role of pili is attachment either to other bacteria during conjugation(see later) or possibly to surfaces during the formation of bacterial films. Recent work indicates that pili may also be used as grappling hooks to aid movement across surfaces.

Nutrition

Some prokaryotes are saprotrophic, some heterotrophic and some are autotrophic. Bacteria and archaea are also to be found in the mouth and digestive tract living in a mutualist association.

The cyanobacteria – often confusingly called blue-green algae (they are not algae at all) are autotrophic. Some use pigments similar to those found in green plants to trap light energy. The pigments are found either bound to the inner face of their membranes or within membrane bound organelles called thylakoids. These are the only membrane bound organelles found in prokaryotes.

Structure/types

Cell shapes – bacterial species have distinctive shapes. There are spherical cocci, rod shaped bacilli, comma shaped vibrio, and spiral shaped spirillum. Some species occur singly, others in twos or even clusters. The diagram below summarises the main configurations.

Incidentally an example of spirillum is *Treponema* (syphilis) and of vibrio species *(cholera)*

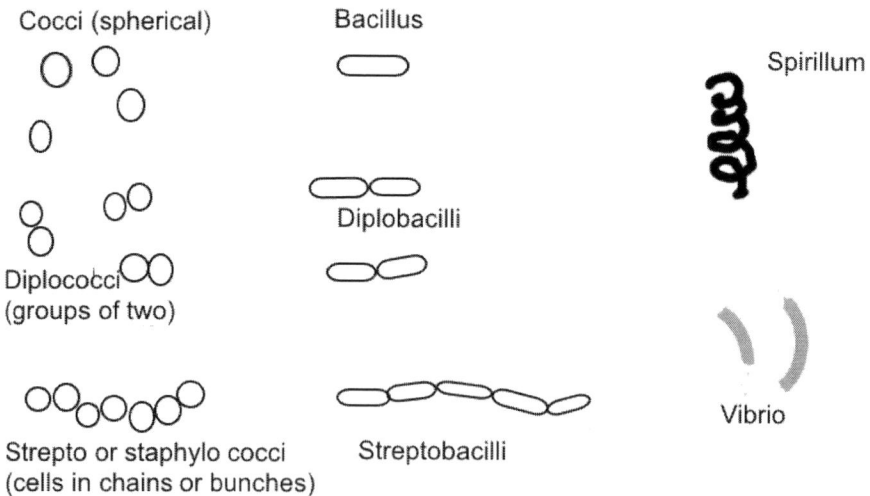

Cocci (spherical)

Bacillus

Spirillum

Diplobacilli

Diplococci
(groups of two)

Strepto or staphylo cocci
(cells in chains or bunches)

Streptobacilli

Vibrio

Apart from the shape of individual bacteria, different species often form distinctive shaped colonies (colony morphology). Colony morphology is another of the microbiologist's tools to aid identification. Another is differential staining, such as Gram's staining.

Examples of infections

We have known of the existence of bacteria for a very long time. In the late 1600s Antonie van Leeuwenhoek an early microscopist recognised them. Even earlier, researchers such as Girolamo Fracastoro had suggested that disease was caused by unseen living particles, but the prevailing view was that disease was caused by 'bad air' (the miasma theory).

John Snow, Ignaz Semelweiss and Joseph Lister were three 19th century surgeons who separately recognised that cleanliness and hygiene especially during operations led to better outcomes for the patient. John Snow's most famous contribution lies outside the operating theatre in his tracing the source of a serious London cholera outbreak to the communal water pump in Broad Street, Soho. The final leap forward came with the 19th century work of Louis Pasteur and Robert Koch. In more recent years surprising links have emerged. For instance ulcers of the digestive system used to be attributed to poor eating habits, stress, lifestyle, all of which can play a part. But in the 1980's a revolutionary study showed that an even more important factor was infection by a gram-negative bacterium called *Helicobacter pylori*. (Warren & Marshall, 1983) Unlike a lot of other bacterial species, this little bug could thrive in the highly acid stomach conditions. Thus, it is routine now to treat ulcer cases with antibiotics in addition to healing medications.

Likewise, heart disease has long been known to be a multifactorial condition. There is anecdotal evidence that Victorian dentists linked periodontal disease to an increased risk of developing rheumatic fever (a form of heart disease). More recent research has supported (but not conclusively established) the link. While streptococcal species are found in some patients who have experienced heart attacks it is still not always clear if they were the cause or are just present in many of the population anyway.

In many cases, health problems involving bacterial infection are clear cut – we either get better ourselves, or a course of antibiotics helps.

 Finally, the jury is still out on a link between archaea and disease. In the 1980's a paper suggested a link between Crohn's disease and archaean infection. (McKay, et al., 1985). However more recent work reveals that we still have some way to go in clarifying the nature of any link. (Weinberger & Gophna, 2015).

In the past few years a phenomenon called quorum sensing has been discovered. It has been noted that bacterial populations enter a host and just tick over until they reach a critical mass. The bacteria then release signalling molecules causing an explosion of numbers with the concomitant illness in the patient (as they are now). Recent research has sought to identify ways of jamming these signals. Promising substances have been noted among the phytochemicals (plant substances). The hope is that the new anti-pathogenic drugs tackling quorum sensing will sidestep the problem of resistance.

Transmission, transfer and movement

Many bacteria have flagellae enabling them to swim. However that does not mean that they can behave like mosquitoes, zooming around from victim to victim. The flagellae are only useful in watery environments. Bacteria are transmitted from person to person in any of three ways:

1.By touching a person or an object with bacterial cells on it (like a door handle or loo flush)

2.Through droplet transmission. Someone near you coughs or sneezes and you inhale the microscopic water droplets containing bacterial cells.

3.From the environment – contaminated food, water (Legionnaire's disease) or soil.

Once inside the body they are free to roam and set up house ready for the big push (see quorum sensing above). Interestingly, a lack of flagellae does not mean that a bacterial species is immobile. Species may adopt other methods of locomotion e.g. gliding in myxobacteria. (Mauriello, et al., 2010). Bacteria have been around a lot longer than we have and still surprise us with what they can do.

Bacteria are superb survivors. Heat is a potent way of killing them – but some species faced with hostile conditions enter into a resting state. They form highly resistant structure called endospores which can last for years, until conditions improve, when they become viable bacteria again. Incidentally, domestic freezing does not kill bacteria; it just slows their metabolism right down. Thus on defrosting, bacterial numbers will be at safe levels – for the time being. However if you wait a day and then change your mind and refreeze the food, bacterial numbers have increased to a higher baseline, creating a very risky situation.

Reproduction

Bacteria reproduce very quickly in ideal conditions. A doubling time of 20 minutes in the laboratory is a working average. However much shorter times have been found in nature- a species in Tanganyika has been found to double in numbers in 10 minutes. (Elsgaard & Prieur, 2011). Much of the time bacteria reproduce by an asexual process called binary fission (the prokaryotic version of mitosis in eukaryotes). The process involves doubling the genetic material then splitting the cell into two.

However bacteria are pretty promiscuous and things get steamy when they indulge in a process called conjugation. The bacteria extend specialised sex pili which act as a linking tube between the two cells. DNA, usually a plasmid, is then transferred and spliced into the DNA of the recipient cell. Thus variation and new genetic combinations can form. Conjugation has been described as the bacterial equivalent of sex.

Because bacterial numbers can grow exponentially, the chances of mutations also increase. This is how antibiotic resistance develops in a species. The antibiotic kills all but a few mutant bacterial cells which are antibiotic resistant. These are now able to grow exponentially without competition from the now dead original population. As noted above, conjugation increases the spread of new genes into the wider bacterial population.

Significance to humans

"Don't forget that the flavours of wine and cheese depend upon the types of infecting microorganisms." (Fischer Martin 1879-1962)

So far you have mainly heard about the downside of bacteria – infection. But without bacteria we would not exist in our present state. Apart from food and drink, we use bacteria for biotechnology. Their accessible genetic material makes them ideal for inserting human genes (e.g. for insulin, growth hormone). We also make extensive use of bacterial enzymes – enzymes from thermophilic bacteria are ideal for our high temperature processes- for instance washing machines and industrial applications.

Over the past 20 years there has been great interest in a bacterial species Agrobacterium tumefaciens. Why? Because it looks like a bacterium but behaves like a virus. This bacterium can infect plant material and insert a section of its own DNA into the plant DNA, causing the growth of galls. Work with Agrobacterium species as vectors of beneficial genes continues. (Gelvin, 2003)

Bacteria are also essential for biogas production. The large scale dairy farming operation run by Wyke Farms in Somerset uses bacteria in a substantial digester plant to convert the energy in cow manure and slurry into power which is then used to drive the cheese making operation. Throughout the UK, methane is tapped off from landfill being digested by bacteria and used to drive generators. Some of the power then passes through the National Grid to domestic and industrial users.

All these are relatively high tech uses. However long before tech even became low tech, bacteria were playing vital roles in the nutrient cycles. The nitrogen cycle for instance relies heavily on soil borne bacteria to produce the nitrates which are then used by plants.

Viruses

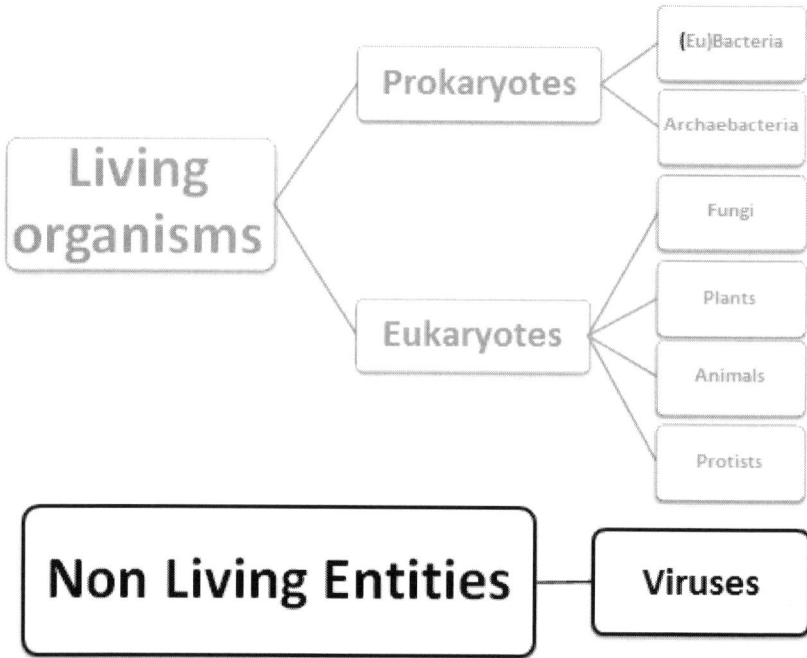

Where are viruses found?

Essentially viruses have to enter living cells in order for more viruses to be assembled and released to infect more cells. Biologists have estimated that there are in excess of 10^{31} different types of virus worldwide - that's 10000000000000000000000000000000! (Zimmer, 2013)

Accordingly most biologists will answer the question "Where?" by stating that anywhere there are living cells, there are viruses to infect them. You will discover why living cells are necessary a little later. Some on the maverick fringe contend that viruses exist regardless of the continuous presence of life. Advocates of the panspermia theory suggest that viruses travelling on comets and meteorites arrived here on Earth several billion years ago and provided the building blocks for all the life that followed.

Presumably on the journey from … where (?), they did not have the company of living cells as fellow travellers.

Unlike bacteria, viruses can not carry out the seven characteristics of life on their own; and they are therefore classed as non-living. It is important to realise that the terms 'living' and 'non-living' are not absolute; they are simply terms invented by people to help categorise the world about us. For now just accept the distinction otherwise you will never get any further!

There is much uncertainty as to which came first – prokaryotes or viruses. There are three suggestions as to how viruses appeared:

1. Viruses are remnants of larger organisms that are now able to exist independently. (The regressive theory).

2. Viruses developed from genetic material that (somehow) was able to move (or be moved) between cells. (The progressive theory).

3. Viruses were in fact the first on the scene when gas ball Earth started forming into a planet. After all they are simple chemical machines even if they behave in puzzling ways. (The virus first theory)

If you are interested in probing deeper, the Nature article by David Weissner should be your first port of call. (Weissner, 2010).

Structure/types

Viruses are tiny; in most cases much smaller than the smallest prokaryote cell. The usual range is **between** 10 – 300nm. **A** nanometre is a thousandth of a micrometre and therefore a millionth of a millimetre. There are just a few exceptions – the giant viruses, such as the Mimivirus, **Pithovirus** and more recently the **Pandora virus**. These can square up to prokaryote cells and measure up to 1.5µm (1500nm).

With 10^{31} different types of virus, categorising them could be a headache! One way is to divide them by whether or not they have an envelope as per the diagram below.

Enveloped and non-enveloped viruses

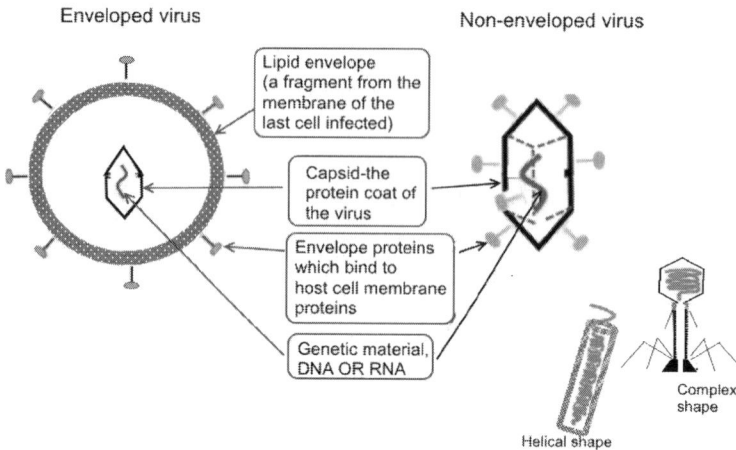

Enveloped virus   Non-enveloped virus

Lipid envelope
(a fragment from the
membrane of the
last cell infected)

Capsid-the
protein coat of
the virus

Envelope proteins
which bind to
host cell membrane
proteins

Genetic material,
DNA OR RNA

Complex
shape

Helical shape

The differences are related to the mode of entry and exit of the virus from the body cells they infect – of which more later. However, presumably this still leaves quadrillions in each of the two categories. We have already seen that viruses vary considerably in size so that is a start. You can also see that viruses differ in shape. The common shapes are icosahedral (the top two) and helical and complex for the bottom two non-enveloped viruses.

So, apart from size, shape and whether they are enveloped or not, another set of distinguishing features concern the genetic material within each type of virus. If you are unfamiliar with the differences between DNA and RNA and the basics of protein synthesis, just try and get hold of the basics. I have shown these in bold. The more complex optional material is in italics.

The possibilities are:

Double stranded DNA viruses (dsDNA)

Single Stranded DNA viruses (ssDNA)

Double stranded RNA viruses (dsRNA)

Single stranded RNA viruses (ssRNA) *comprising the + or positive sense strand (which is basically mRNA). Thus once it enters the cell it causes the cell to start synthesising viral protein rather than host cell protein.*

Single stranded RNA viruses comprising the – or negative sense strand. This can form a template for the synthesis of mRNA within the infected cell.

Retroviruses - are a single strand + RNA virus. However they **act by inducing the infected cell to produce a DNA version of the viral genetic code and then splicing it into the nuclear DNA of the host cell.**

Transmission, transfer and movement

Unlike bacteria, viruses cannot move of their own volition. Effectively they rely on a chance collision between the protein receptors of the virus and the matching receptors on a host cell. The chance collision is facilitated by currents within the host – e.g. airways, digestive movements, blood flow. A successful collision (for the virus) then enables the viral infection cycle to be triggered.

Viruses work by infecting cells, then using the cell machinery to produce multiple virus particles. When the new viruses (or virions) are released, they may wreck the cell in the exit process (burster viruses) or they may leave quietly and leave the virus infected cell intact as an ongoing virus factory (budder viruses).

As a generalisation, the viral infection (not life!) cycle involves:

1. Entry

2. Replication of viral genetic material. (DNA/RNA replication)

3. Production of multiple virions by the infected cell. (Viral replication)

4. Release of the new virions so they can collide with and infect other cells.

There are a lot of variants of the way each stage is completed by different viruses, but here is a potted summary.

Viral entry

The binding of the virus-host cell receptors then allows one of three **entry routes** to be activated:

1.Fusion. This is a feature of enveloped viruses. The envelope is made of the same substances as the host cell membrane (i.e. a phospholipid bilayer). Thus once binding occurs the two membranes fuse and merge, allowing the viral capsid to enter (the viral envelope is now incorporated into the host cell membrane).

2. Endocytosis- as an alternative to 1. Binding may trick the host cell into obligingly conveying the virus into its cytoplasm by a process called endocytosis. This obviously works well for non-enveloped viruses; however some enveloped viruses such as Hepatitis C enter by the endocytosis route.

3. Injection. Bacteriophages are viruses that infect bacteria. A well-studied example is the T2 phage which binds, then injects the viral genetic material through a puncture in the host membrane.

Replication of genetic material

1. Before the viral DNA or RNA can replicate, it has to be released from its capsid suit of armour. This is a process called 'uncoating'.

2. What happens next varies hugely and is *in general* related to the type of genetic material. Viral replication depends on a copy of viral RNA (actually mRNA produced by reading the relevant DNA code) being present. As noted above, genetic material of ssRNA (+) viruses is already at this stage, though some viruses will still trigger the ssRNA to replicate, presumably increasing the ultimate rate of virus production. The DNA/RNA of other viruses will often (but not always) be replicated within the nuclear area.

Production of multiple virions by the infected cell

The viral DNA may then be spliced into the nuclear DNA of the host cell using enzymes called integrase enzymes. There it will remain – for ever and be passed on to the next generations – albeit in a non-pathogenic form (as far as we know). In order for virus manufacture to commence, a RNA copy of the new nuclear virus gene is made, using the corrupted DNA as a template. This copy is called mRNA (messenger RNA). It is carried out of the nucleus to the ribosomes, which are the workbench for protein manufacture.

Release of the new virions so they can collide with and infect further cells

1. Enveloped viruses are in capsid form within the host cell cytoplasm, but as they exit the cell they acquire a wrapper of host cell membrane, which becomes their envelope. This process may

leave the infected cell alive, or in a weakened state, or dead. This process is called 'budding'.

2. Non-enveloped viruses are called 'bursters' as they burst out of the infected cell in a much noisier and more destructive manner. Usually the host cell will be killed by the exit of the newly made virions.

Significance to humans

Viruses are significant causes of disease in humans. One of the most exciting early links involved the Epstein-Barr virus. There are several exciting accounts of this discovery - doggedly resisted by the scientific establishment for many years, in book form.

It seems reasonable to us now that anything that interferes with the integrity of our nuclear DNA may also disrupt the normal controls on cell division, leading to an increased risk of cancer. But not just cancer - the damage caused by viruses may be at the root of a wide range of conditions. Increasingly the role of enteroviruses is being investigated in such conditions as diabetes, heart problems and the inability of couples to conceive.

Of course, many viruses are not pathogenic to man. They may however cause crippling losses economically and socially by affecting the food production system. In many cases they are transmitted by sap feeding insects.

However, viruses are also essential to humans. Some viruses help in fighting off pathogenic bacteria, others can counteract the effects of loss of our intestinal gut flora. (Roossink, 2015).

Viruses have also been used to piggyback substances or genes into diseased cells. The brain is notoriously difficult to get medicines into due to the blood-brain barrier. Work with specially weakened viruses – notably the adenoviruses has shown great promise for the future.

Fungi

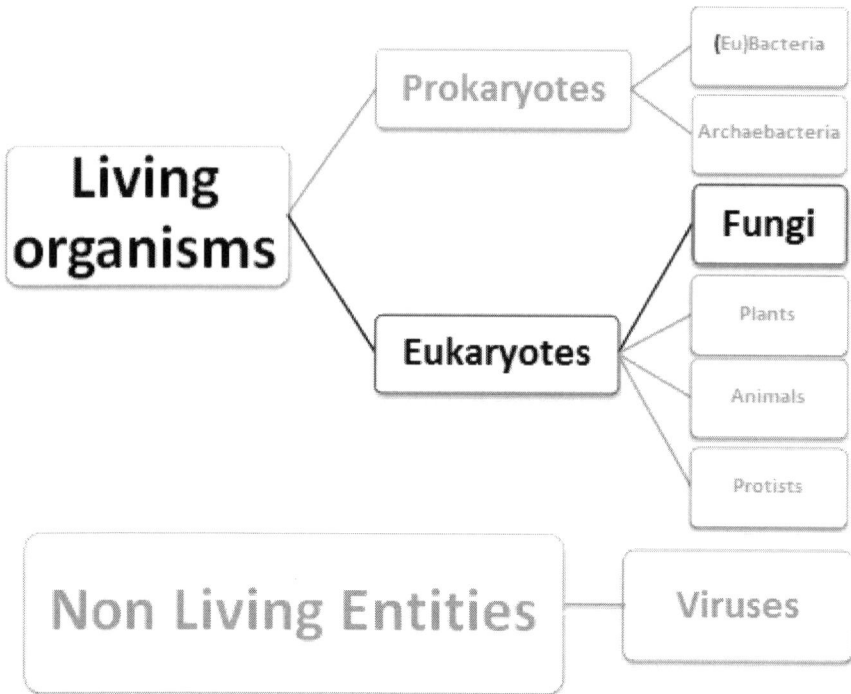

Where are fungi found?

Fungi are found in soil, fresh and saltwater, air and in and on other organisms. Some are beneficial, some harmful, some downright dangerous.

The fungi kingdom is a huge one – estimates put the number of species in excess of 1.5×10^6 (that's 1.5 million in the old money). (Moore, et al., 2011). It is a hugely diverse and complex amalgam of organisms and the classification system seems to change daily. Search the literature and you will find current phyla of fungi put at 3,4,5,6,7 ..... For that reason this book will keep it simple and start by dividing the types into the filamentous and non-filamentous species.

Because they do not have skeletons, dating the emergence of fungi by fossil evidence is tricky, but the oldest fossil evidence found so far dates back to four hundred and sixty million years ago (4.6×10^8 Ma)

One thing all fungi have in common is that they are heterotrophs and they digest their food externally. Thus they are chemo-organoheterotrophic, deriving their nutrients from a wide variety of substrates. There are 3 feeding types:

1. Saprotrophs – these are the decomposers and form the majority.

2. Necrotrophs – These are pathogenic. They infect host tissue (usually plants), kill it then digest it and absorb it.

3. Biotrophs – also infect organisms, mainly plants. An example is powdery mildew. Like Necrotrophs these grow on plants but do not kill as quickly.

Most fungi are aerobic – but some (e.g. the group called Chytrids) are anaerobic. These can be found deep in the digestive systems of many animals-especially ruminants such as cows.

Fungi are very robust organisms, and occupy regions of the world with a wide range of temperatures. The three categories by temperature tolerance are as follows.

1. Mesophiles – these flourish between temperatures of 18-22°C and form the majority of fungal species.

2. Psychrophiles – these can grow at low temperatures of less than 10°C. Interestingly, freezing does not kill fungi. In fact a research study found that most fungi continued growing, albeit slowly at 0°C (Flanagan, 1978)

3. Thermophiles – tolerate hot conditions well and are found at temperatures of 37°C and above.

This obviously has implications for humans – e.g. in the field of food storage and safety. A further concern involves the presence of potentially pathogenic fungi in water distribution systems. (DEFRA, 2011)

It has also been found that certain fungi can be resistant to certain disinfectants, so choosing the right cleaning or sterilising agent is crucial. Again this is of particular concern in food production. (Bundegard-Nielsen & Nielsen, 1996)

When fungi are exposed to unfavourable conditions, they produce thick walled survival spores called memnospores. Most textbooks will tell you that these can survive for a long (unspecified) time. Theoretically this is true and the spores will certainly be resistant to adverse climatic conditions. However survival is markedly reduced by the action of other microorganisms which feed on and destroy the spores. Often fungal survival is better assured through the production of sclerotia. (Dix & Webster, 1995). Incidentally a fungal parasite of the cereal crop Rye produces sclerotia containing a toxin which can be fatal to humans. Interestingly, substances within these same sclerotia have also been extracted for use as drugs – notably to facilitate childbirth in humans!

Structure/types

Two basic types of fungi are the filamentous type (e.g. *Rhizopus*-bread mould) and the spherical/semi-spherical groups (e.g. yeasts *Saccharomyces cerevisiae*)

Rhizopus – a filamentous fungus

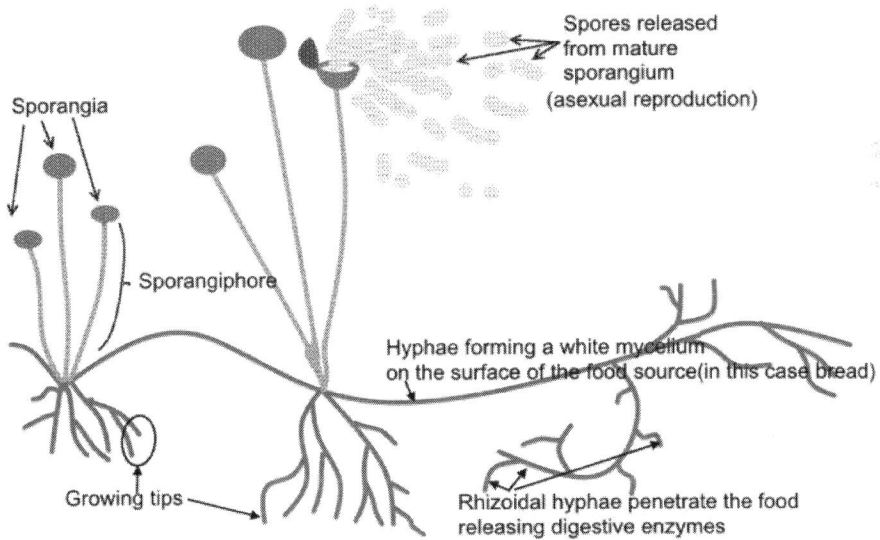

Sporangia

Spores released from mature sporangium (asexual reproduction)

Sporangiphore

Hyphae forming a white mycelium on the surface of the food source(in this case bread)

Growing tips

Rhizoidal hyphae penetrate the food releasing digestive enzymes

Yeast (*Saccharomyces* spp) a spherical fungus

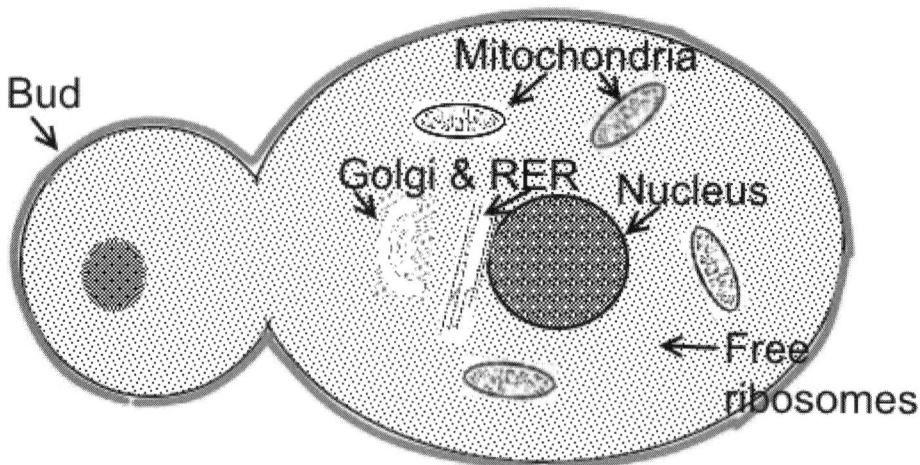

Notice that all fungi except mycoplasmas possess a cell wall containing chitin outside the membrane.

Some fungi – especially pathogenic ones - are dimorphic and exploit both filamentous and spherical forms at different stages of their life cycle. The spherical form is very good for conduction through fluids in plants or animals; i.e. blood, lymph or fluids in plant xylem and phloem. Once the pathogenic fungus reaches solid target tissues, the fungus develops the mycelial (filamentous) form to begin feeding, destruction and, worst case scenario, death of the host.

A little earlier there was mention of anaerobic species of fungi (within the group called the Chytrids). If you have read about cell respiration in Book 1 of this series (see Further reading, later) you can probably see that this should limit the release of energy (in the form of ATP) within cells to around 5%. Not so – these little fungi have specialised organelles called hydrogenosomes which can maximise energy release even anaerobically. Hydrogenosomes are similar to mitochondria but there are key differences – they lack DNA for instance. It is tentatively suggested that both mitochondria and hydrogenosomes evolved from the same common ancestor.

No discussion of the sometimes weird life within the fungi kingdom would be complete without mention of lichens. Lichens are comprised of a fungus and an algae (or sometimes a cyanobacterium) living in close association. Fossil remains date from 400Ma. Lichens tolerate very wide temperature extremes, as well as water availability. They are a pioneer species. In some parts of the world they are a valued food source – for instance forming 95% of the reindeer diet.

Lichens vary both in *form* and *colour*. In terms of form, lichens may be:-

- Crustose. These are the flat crust like lichens you commonly see on roofs, tree trunks, stone walls etc. In the UK these probably exhibit the greatest range of *colour*.
- Foliose. These are described as 'leaf like'. Essentially this means the outer edges are not in intimate contact with the substrate. Some foliose lichens are lobed whereas others are circular (attached by a single central cord). Some foliose lichens can feel quite leathery.
- Squamulose. (Think of squamous epithelium) The appearance of this form is in between the foliose and

crustose type. Thin and crust like but often with small thin lobes. Not leathery.

- Fruticose. Possessing erect structures-sometimes finely branched. Fruiting bodies are at the end of these. Some *cladonia* species have beautiful red tips.
- Gelatinous.

Identifying lichens can be extremely tricky!

Colour

Lichens range through shades of green, through blue green to yellow, orange and silvery blue green. If you have studied photosynthesis you may have carried out **chromatography** of leaf pigments. You may recall the action and absorption spectra. You may also recall that different pigments are excited at different wavelengths of light. Many studies have been carried out to investigate the relationship between position and orientation of lichens with their colour.

Lichens generally grow very slowly. They are also long lived - some lichens have been dated at 4500 years old. However you must be aware that man has not always shared an interest in lichens... in some churches people become almost paranoid about removing lichens from stonework and gravestones. (they are usually the same people who want to 'remove' bats because they cause 'mess')

Reproduction in lichens can be complicated- both sexual and non-sexual reproduction occurs.

Lichens do not have true roots- they are like mosses and liverworts in this respect. However whereas the latter two are attached to the substrate by rhizoids, some lichens are attached by *rhizines* based on a bundle of hyphae.

Fungi and infections

There are two broad headings of fungal disease in humans. (1) Skin mycoses and (2) Systemic mycoses.

(1) **Skin mycoses** may infect living skin - especially in the tropical regions of the world. As many of these pathogenic fungi are soil dwelling species, people walking barefoot are particularly likely to become infected. Quite a few demonstrate the dimorphism mentioned earlier. These are known as superficial mycoses. Some fungi however cause disease in non-living tissues such as keratinised skin, hair, nails. Athlete's Foot and Ringworm are well known examples of these cutaneous mycoses which are very common. Cutaneous mycoses may be:

(i) geophilic – soil dwelling fungi which can infect through contact with soil.

(ii) zoophilic – typically resident on animals but readily transmitted to humans through touch and handling.

(iii) anthropophilic – these rarely infect other animals but are pretty specific to humans. Ringworm is an example.

2. **Systemic mycoses** are often due to primary opportunistic pathogens. They may be inhaled as spores which then spread from the lungs to the rest of the body. *Cryptococcosis* is found and spread via pigeon droppings, in humans it can lead to fatal meningitis. Opportunist fungi are more likely in the immunocompromised and in subjects undergoing modern treatment- e.g. for cancer. Anything that affects the normal bacterial flora of the body can facilitate fungal infection. That is because normally our resident bacteria reduce the ability of fungi (e.g. *candida*) to attach to body surfaces.

Spore inhalation can also trigger allergic responses, leading to conditions such as allergic aspergillosis (the fungus is aspergillus). However we are all inhaling spores daily from the environment – our immune systems deal with this routinely without any fuss. It is primarily those with high occupational exposure who are at risk of developing problems.

Other problems with fungi - Fungi can cause serious economic losses in food production. Stored food is likely to fall prey to fungal moulds the moment there is enough moisture present. Cultivated crops of all types, cereals, vegetables and fruits can all be ruined by fungal attack. Probably the most famous in the UK is the potato blight famine in Ireland in the mid-1800s, which accounted for a million deaths due to starvation.

Reproduction

In animals and higher plants, most of the time the individuals are diploid, but they produce short lived gametes (e.g. sperms or eggs in humans) that are haploid. Thus most of their lives are spent in the diploid state.

Most species of fungi are different and spend most of their time in the haploid state. The diagrams that follow summarise what happens. As always, there are a few species of fungi which do not follow this.

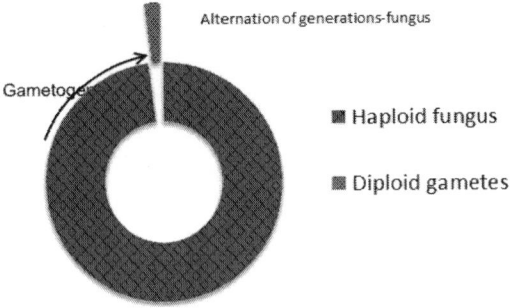

Alternation of generations-fungus

Gametogenesis

■ Haploid fungus

■ Diploid gametes

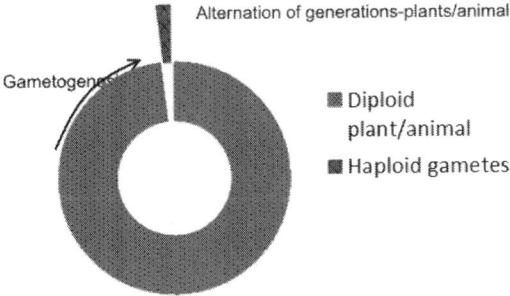

Alternation of generations-plants/animals

Gametogenesis

■ Diploid plant/animal

■ Haploid gametes

Before you throw your hands up in annoyance and decide to give up on fungi, be reassured that most mycelia are definitely haploid!

Sexual and asexual (including vegetative) reproduction all occur in fungi. Yet again we find fungi adopting some very complex strategies. Here is a roundup.

Sexual reproduction.

Among the filamentous species, some species are self-fertile, others are self-sterile.

Yeasts adopt a strategy called 'gene switching'. Yeast is heterothallic, meaning that the same individual can produce cells with either of the alleles A or a. Only A and a can mate. A and A or a and a are not viable to produce new offspring.

Asexual and vegetative reproduction

Apart from sexual reproduction, fungi can also reproduce (and produce clones) as follows:

1. Asexually by producing haploid spores.

2. Vegetatively by new individuals budding from the parent or splitting (fission). Yeasts are a good example.

3. Vegetatively by fragmentation. Fragments of the hyphae are separated from the mycelial parent and start growing to produce a new individual.

Significance to humans

So far you have read a succession of gloomy titbits about the problems and diseases fungi can cause. However fungi, like other microbes (arguably more than them) offer a huge range of benefits to mankind and indeed the whole kingdom of life. Here is a short roundup:

1. **Decomposers** – fungi play a crucial role in the nutrient cycles. They efficiently break down plant (and animal) remains releasing the nutrients back into the ecosystem.

2. **Maximising plant growth** – certain fungi can colonise the root of plants and form a symbiotic relationship (known as a mycorrhizal association). The fungi make essential nutrients available for the plant, which is highly beneficial in poor or sterile soils. These associations may be specific between a certain fungus and a certain plant (ectomycorrhizal associations) or not very specific (endomycorrhizal associations).

3. Fungi as food –

(i) Edible macrofungi (mushrooms) are highly prized by many. The nutrient content varies quite widely, but they can be an extremely useful source of B vitamins for vegetarians and vegans. This is especially so for vitamins such as B_{12} which are otherwise only available in meat.

(ii)Fungi in food processing - Yeast has been used for many thousands of years. The early Egyptians are said to have used it for leavening (rising) bread around 5000 years ago. Baking and alcohol production followed. There is a whole host of food stuffs treated with fungi to enhance their organoleptic qualities ranging from cheeses, Tempe, Soy Sauce, bean curd …. It's a long list!

4. Pharmaceuticals – the discovery of penicillin (from the penicillium fungus) is now in folklore. Other fungi produce other antibiotics (e.g. cephalosporins, cyclosporins). Apart from being a powerful antibiotic, the some of the cyclosporins are now used as immunosuppressants. There are many other uses for fungi, including very promising discoveries that can be harnessed in the treatment of certain cancers.

One reason that fungi have proved so useful to humans in the fields of biotechnology is that they are also eukaryotes and so possess similar cellular machinery to ours.

5. Chemical production – As with bacteria, thermophilic species of fungus possess enzymes that are not denatured readily at high temperatures. These enzymes are ideal for industrial (and domestic) applications involving high temperatures. Over the past 20-30 years there has been increased interest in bioremediation as a way of tackling pollution problems.

Immunity and anti-microbial medications

Introduction to Immunity and anti-microbial medications

Despite living in the close company of trillions of pathogenic micro-organisms the number of times we fall seriously ill is mercifully few. In this book you are introduced to what we currently know about the working of the incredible immune system that keeps us disease free for much of the time. The book is focused on infectious diseases. You will also discover some of the drug treatments that can be used in order to help us when the going gets tough for our immune system.

In order to get the most from this book you need to know the basics of cell structure in eukaryotes and prokaryotes. (The book Basic Introductions to Biology (1) - Cells Tissues and Circulation by Pip Flowers gives a concise introduction)

I hope you enjoy this introduction to immunology and the medications of infectious disease.

Pip Flowers

Summer 2017

Types of Disease

Disease is the term used to describe conditions affecting the proper functioning of body organs and/or body systems. The terms illness and disease are often interchangeably used. Strictly speaking however, the term illness describes a *perception* experienced by the sufferer. It may be due to disease, or it could be due to for instance, trauma.

Broad classifications of disease are shown in the following table.

Broad classifications of disease.

Disease may be	Types	Examples
Infectious	Bacterial	Streptococcal(sore throats) E coli (sickness/diarrhoea) Mycobacterial (TB) etc.
	Viral	The flu and colds Rotaviruses(sickness/diarrhoea) Herpes (cold sores amongst others) etc.
	Algal	Protothecosis Shellfish poisoning(strictly speaking poisoning by toxins rather than infection)
	Fungal	Candida (Thrush) Trichophyton (athletes' foot)
	Protozoa	Plasmodium(malaria) Entamoeba(bloody diarrhoea, sickness, death) Giardia(diarrhoea, sickness, stomach cramps
Non-infectious	Inherited	Cystic fibrosis PKU Predispositions to diseases-e.g. some cancers
	Auto-immune	Where the body's immune system starts attacking your own healthy tissue Eg: Rheumatoid arthritis Glomerulonephritis Type I diabetes
	Environmentally triggered	Systemic sclerosis Hay fever Emphysema Cancers (e.g. resulting from radiation exposure, asbestosis)

The more we know about disease and the human body, the more we are finding some blurring of these nice tidy boundaries. For instance viruses are now implicated strongly in certain cancers, bacteria in certain forms of heart disease as well as stomach ulcers. Autoimmune diseases (e.g. glomerulonephritis) may often be set off by an infection. Conditions such as coronary heart disease are termed multifactorial- in other words you get it when several risk factors come together.

Effects of microbial infection.

Microbes can cause harm by:

1. Destroying tissue (e.g. the notorious flesh eating bacteria causing Necrotizing Fasciitis)
2. Releasing toxins. They may release these as part of their normal activities (exotoxins) or may contain endotoxins which are released when the bacterium is destroyed (e.g. by stomach acids),
3. Subverting normal cell function. For instance viruses hijack our nuclear DNA and turn our cells into virus factories. Other microbes affect membrane pumps or the cytoskeleton.

Over the past few years it has been discovered that many infective bacteria 'talk' to each other. When they enter the body they may just tick over, multiplying gradually until they reach a critical mass, at which point cell signalling between them triggers an explosion of growth and the symptoms of infection. This communication is known as 'quorum sensing'.

The body doesn't sit meekly by allowing these microbes free reign. The immune system has a whole battery of responses, some of which will be outlined in the rest of this book.

In some circumstances clinical interventions may be appropriate. Later in the book you will learn about these.

Defence ……. and immunity

Life was so much easier in the good old days. You could pigeonhole the body's systems conveniently and tidily. The immune system was visualised as a series of barriers that a pathogen had to surmount in order to infect the host.

However the more we learn about the human body and especially the immune system, the more blurred the 'boundaries' become. In fact we have now reached the stage where it is no longer appropriate to talk of boundaries at all. Instead we see integration. The emphasis switched from a spatial view of the immune system to a functional view.

The development of this view has been:

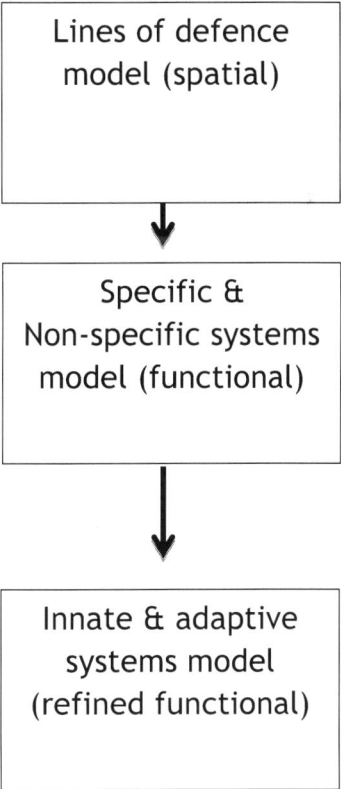

```
┌─────────────────────────┐
│    Lines of defence     │
│    model (spatial)      │
│                         │
└─────────────────────────┘
            │
            ▼
┌─────────────────────────┐
│       Specific &        │
│  Non-specific systems   │
│    model (functional)   │
│                         │
└─────────────────────────┘
            │
            ▼
┌─────────────────────────┐
│    Innate & adaptive    │
│     systems model       │
│   (refined functional)  │
│                         │
└─────────────────────────┘
```

The lines of defence model is still used in many texts and can be a useful way of visualising what is happening at a very basic level. In this chapter the first line of defence is described; but be aware that different books may differ slightly in their classification. Bear in mind too, that defence implies something is always in place whatever the immediate risks; it is different from an attack response (i.e. immunity)

Lines of defence model

The first line of defence against pathogens includes:

Cilia and mucus lining the respiratory tract- the trachea (windpipe) and bronchi contain specialised cells. Some (goblet or mucus cells) produce a continuous layer of sticky runny mucus which coats the respiratory passages. Pathogenic organisms and harmful particles stick to the mucus and very few reach the delicate alveoli inside the lungs. Others (ciliated epithelial cells) have cilia projecting from their cell membranes. The cilia create a wave like movement (imagine a Mexican wave) which carries the mucus up to the throat where it is swallowed and then destroyed in the stomach. You will sometimes see this termed the 'ciliatory escalator'.

Ear wax – all too often treated as an annoyance to be removed, ear wax is a self-lubricating, self-cleaning anti-pathogenic substance containing lysozyme, fatty acids and an acid pH all of which create unfriendly conditions for bacteria and fungi intent on setting up shop in the warmth of the ear canals.

Saliva – Saliva contains the powerful antibacterial agent lysozyme.

Skin – Forms a physical barrier preventing microbes entering the bloodstream. Sweat contains salt which can affect osmotic balance of bacteria, also contains an antibacterial enzyme called lysozyme. The skin cells can also produce antibacterial substances called 'defensins'. Plants also produce defensins. More about defensins in the antibiotic section. The pH of skin and also the natural flora may create hostile conditions for pathogens.

Stomach acids – stomach juices in general are very acidic (during digestion) and also contain protease enzymes (i.e. pepsin). Thus any microbes reaching the stomach are likely to be destroyed. Bacterial cell walls and membranes contain proteins, thus they are likely to be broken down very rapidly by enzymes or denatured by the acid conditions.

Tears – Despite the delicate surfaces of the eyeball being exposed all day to irritants and pathogens, we seldom experience eye problems. This is due to tears which contain a battery of highly effective protective substances including lysozyme and surfactant protein-D.

Blood clotting – Some biologists argue that clotting should be included as it is a response to some form of physical damage. Blood clotting is described in the next chapter.

The cat among the pigeons – the complement system.

The discovery of the complement system further confused those nice tidy classifications of the defence and immune system. The blood plasma contains a huge number of different chemicals called plasma proteins. The majority of plasma proteins are made in the liver. These plasma proteins are circulating round the body all the time in an inactive state – until a trigger event happens. In terms of the complement system, the something could be the arrival of a pathogen in the body. This causes the activation of the complement proteins, leading to pathogen destruction.

Thus the complement system acts like a first line defence-but responds like a second or third line defence.

Moral – don't use the lines of defence model too rigidly.

Blood clotting – defence or response?

Clotting serves two purposes. It prevents loss of vital body fluids and it also reduces the chances of pathogens entering the body. Following a wound, surface vasoconstriction reduces blood flow to the wounded area.

Meanwhile *platelets* – normally inert cell fragments produced by the red bone marrow- become sticky. They aggregate (clump together) and are swept to the wound exit point forming a temporary plug or clot. Platelets also release chemicals (cofactors) which help activate blood clotting factors involved in the following processes.

At the same time, the blood clotting cascade is activated. Two inactive plasma proteins called prothrombin and fibrinogen are produced by the liver. The blood clotting factors (see above) cleave prothrombin into the active enzyme thrombin. Thrombin then converts the soluble fibrinogen into insoluble fibrin. Fibrin forms a mesh of protein fibres over the wound, entangling platelets, blood cells and protein into a more permanent insoluble clot.

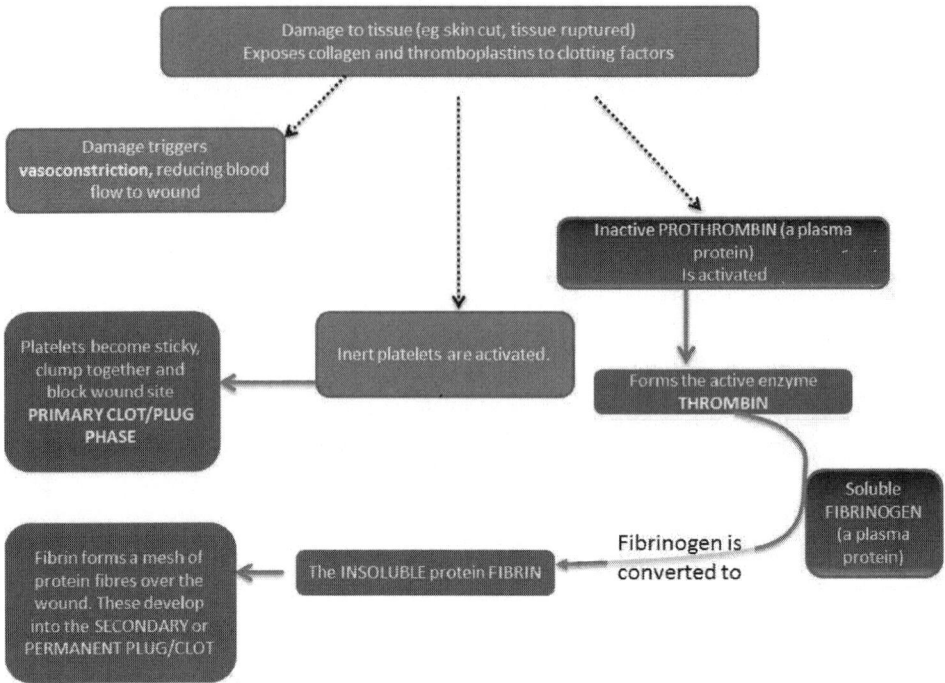

There are of course considerably more substances involved than the diagram shows. However the image and text above give you a good starting point at this stage of your studies. It is important to realise that all this happens simultaneously.

Defence and immunity – immune responses introduced

Under the lines of defence model, the second and third lines of defence involve white blood cells. Although there are many sub types of leucocyte (Eng.) or leukocyte (US), here are the main types.

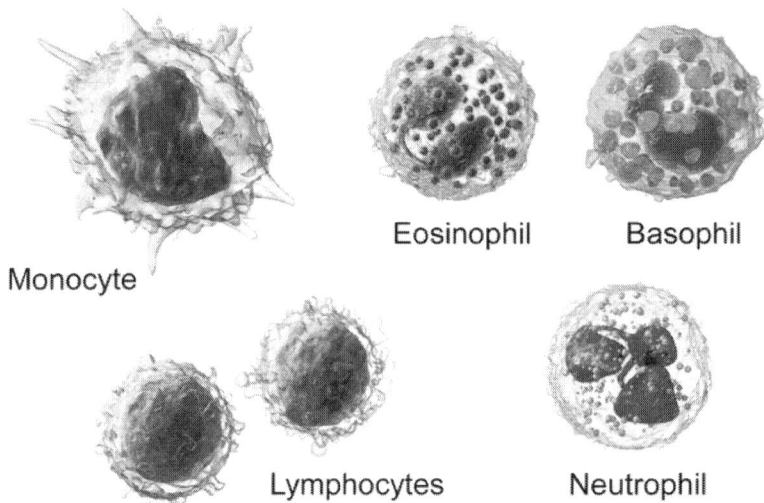

Monocyte

Eosinophil   Basophil

Lymphocytes   Neutrophil

White Blood Cells

Blausen.com staff. "Blausen gallery 2014". Wikiversity Journal of Medicine. DOI:10.15347/wjm/2014.010.
ISSN 20018762. - Own work, CC BY 3.0, https://commons.wikimedia.org/w/index.php?curid=28223981

Numerically and functionally their roles are as follows.

Leucocyte type	Approximate proportion of all white blood cells (%)	Functions
Neutrophils	62	Phagocytic, highly mobile cells. Note the 'lobe shaped' nucleus.
Monocytes	5	Also phagocytic but larger and less mobile. Settle in tissues and are known as 'macrophages'.
Lymphocytes	30	Specialised cells which are part of the adaptive response. Two main types, B lymphocytes and T lymphocytes.
Basophils	≈1	Release chemicals which act as signalling molecules (e.g. cytokines) and also inflammatory chemicals (e.g. histamine)
Eosinophils	≈2	Have some phagocytic activity. Also important in killing blood borne parasites. Eosinophils are important in balancing the activities of basophils. For instance they may release histaminase to modulate the effects of histamine release by basophils.

Similar to the monocytes are the three types of **dendritic cell**. These will not be discussed in detail beyond saying that they form a vital link between the innate and adaptive systems and play a part in antigen presentation, described later.

The inflammation response

Inflammation can be triggered by a number of events. These include injury, infection, exposure to allergens or irritants etc.

Here is the sequence of events following a puncture wound to the skin.

	An 'insult' occurs to the body. This could be an injury, allergen etc. In this case bacteria have entered through a puncture wound.
	The **mast cells** in the tissues of the dermis release the chemical **histamine**. Histamine makes capillaries more 'leaky' and increases flow of blood carrying neutrophils to the wound site. The increased flow of blood brings extra heat and redness to the area. The increased porosity of the capillaries allows more tissue fluid to ooze out (swelling) which presses on pain receptors causing pain.

49

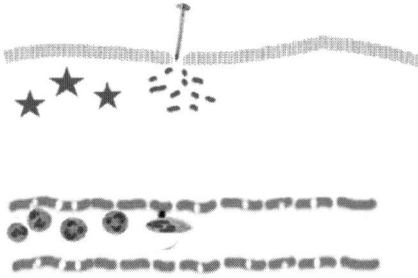	**Adhesion molecules** form on the capillary endothelium forming a brake mechanism on the neutrophils as they roll past in the bloodstream. The neutrophils also start to flatten.
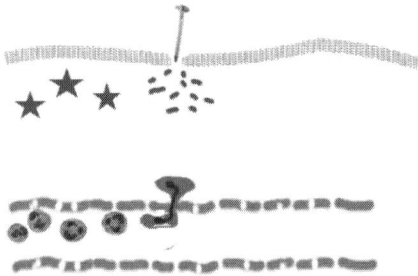	Due to their lobed nucleus (think of a string of sausages), the neutrophils extend and flow through the openings in the capillary wall. This process is called **diapedesis**.
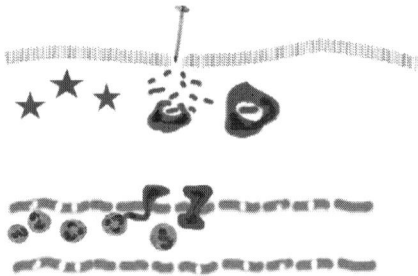	Neutrophils are attracted towards bacterial chemical wastes and toxins by a process called 'chemotaxis'. Neutrophils then flow round the bacterial cells enclosing them in a vesicle made up of neutrophil membrane. The vesicle is called a **phagosome**. Lysosomes then attach to the phagosome and inject destructive chemicals into it, destroying the bacterium.

The adaptive response- lymphocytes.

This is a highly specific response, formerly labelled as the third line of defence. It involves lymphocytes.

Type B lymphocytes – are made in the **b**one marrow and produce anti**b**odies. This is also called humoral immunity.

Type T lymphocytes – come from the **t**hymus gland and give rise to the cy**t**otoxic response or as it is also called, the cell mediated (or cellular) response.

Antigens are glycoprotein markers on the outer face of cell membranes. They are distinctive to each individual and act like a fingerprint. Antigens help the immune system distinguish between 'self' (therefore safe) and 'non-self' (potentially dangerous) cells... Antigens on pathogens alert leucocytes to the need to mount a response.

Type B (humoral or antibody response)-main stages
1. Recognition and binding – receptors on the lymphocyte membrane will bind to complementary shaped antigens on a pathogen.
2. The binding triggers the B lymphocyte to go into very rapid cell division (mitosis) to give rise to a large number of specialised plasma or effector cells plus a memory cell.
3. The plasma cells are optimised for rapid protein synthesis and produce millions of protein antibody molecules.
4. The antibody molecules bind strongly to the bacterial antigens leading to their destruction.

Some more detail about the B lymphocyte response.

1. The lymphocyte receptors in stage 1 (above) are actually antibody molecules produced to deal with previous pathogens. If the same pathogen has been encountered before, the fit with the fork of the 'Y' shape will be extremely tight and the antibody molecules produced via memory cells will be highly effective from the outset. If this specific pathogen has not been met before but something with similar antigens has, there will be Type B lymphocytes with receptors that are similar and fit the fork more loosely. The immune system then undergoes a rapid period of experimentation until the ideal antibodies are produced.
2. The plasma or effector cells are optimised for protein production and they do not have antigen receptors.
4. Antibodies may immobilise the pathogens by binding to them or they may lyse the bacterial membrane.
5. When antibodies bind they do so at the 'fork' end. The single or 'constant region' end acts as a beacon to attract phagocytic cells which swarm to the battlefield and destroy anything marked with antibody. Marking by antibody is an example of 'opsonisation'.
6. The base of the fork acts like a hinge, enabling antibodies to bind with more than one antigen.

Summary of the B cell antibody response.

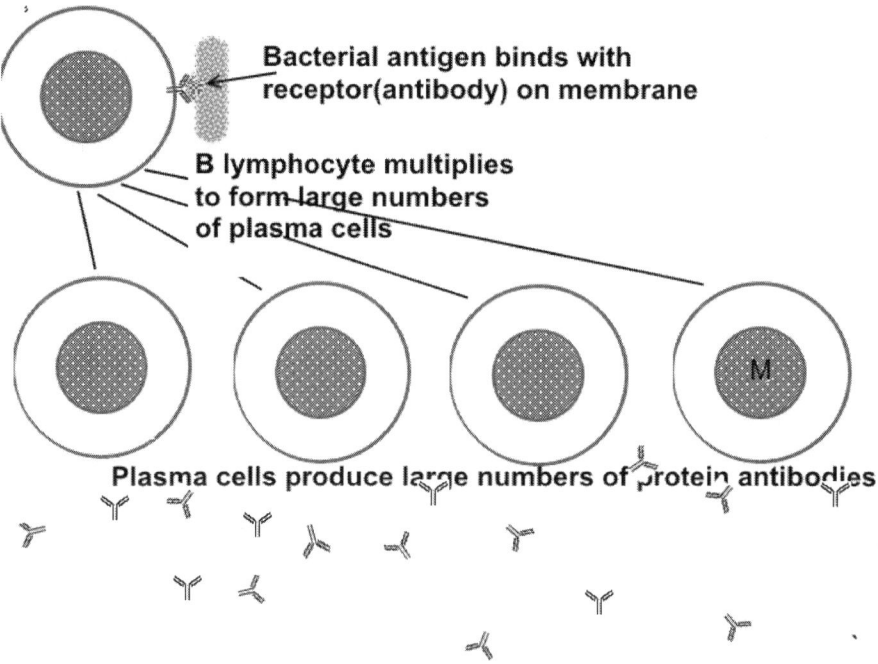

Bacterial antigen binds with receptor(antibody) on membrane

B lymphocyte multiplies to form large numbers of plasma cells

Plasma cells produce large numbers of protein antibodies

Structure of an antibody molecule showing the variable (antigen binding) ends.

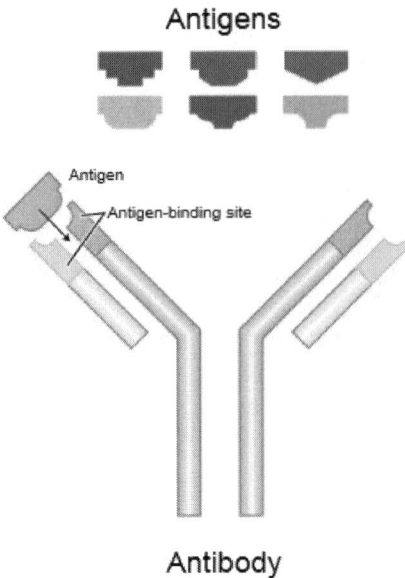

Antigens

Antigen

Antigen-binding site

Antibody

Type T (cell mediated or cytotoxic response) – main stages

Whereas the antibody response works well with circulating pathogens (e.g. bacteria), you may recall from the previous section that viruses pose a different problem. The problem arises because viruses enter cells and are therefore inaccessible to antibody molecules while they are replicating. Fortunately viral attack leaves a viral footprint on the host cell membranes in the form of viral antigen. This is enough for T-lymphocytes to mount a response.

Recognition and binding – receptors on the lymphocyte membrane will bind to complementary shaped (viral) antigens on a body cell that has been infected by virus...

The binding triggers the T lymphocyte to go into very rapid cell division (mitosis) to give rise to a large number of T-killer cells.

The T-killer cells bind to any body cell displaying viral antigen and destroy it, along with the virus it harbours.

Some more detail

The lymphocyte receptors in stage 1 (above) are called T-cell receptors (TCR's). Unlike B lymphocytes, they are not antibody molecules. However they show a very similar specificity to antigens.

The T-killer cells are optimised for destruction of infected cells, thus they have the same TCR's as the original T cell...

Destruction by T-killer cells may be through release of chemicals e.g. perforins) or through the induction of apoptosis by the infected cells.

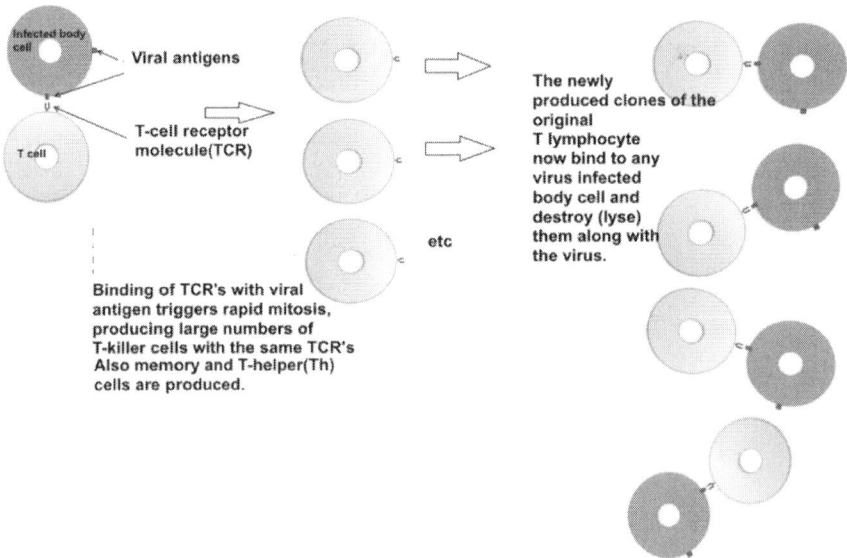

Infected body cell

Viral antigens

T-cell receptor molecule(TCR)

T cell

Binding of TCR's with viral antigen triggers rapid mitosis, producing large numbers of T-killer cells with the same TCR's Also memory and T-helper(Th) cells are produced.

etc

The newly produced clones of the original T lymphocyte now bind to any virus infected body cell and destroy (lyse) them along with the virus.

Lymphocyte responses – going deeper.

Here are some further notes about the lymphocyte responses.

1. T-lymphocytes mature in the thymus gland (unlike B cells which mature I the bone marrow). T lymphocytes respond to infection in a very aggressive manner and it is vital that they correctly distinguish self and non-self. In the thymus gland, new T lymphocytes are 'bench tested' before release into the bloodstream. Any T cells which attach self-tissue are destroyed. The destruction is actually cell suicide -the faulty T cells are induced to undergo apoptosis. Phagocytic cells then complete removal of the debris.

2. Th1 and Th2 – two types of T helper cell are produced. It is still not entirely clear what each do but in general terms,

    (i) Th1 are effective in the fight against bacteria, viruses and parasitic worms (e.g. helminths) and in promoting the cytotoxic T cell response.
    (ii) Th2 are effective in the release of chemicals (called cytokines) which activate a B cell response

3. In fact the Th cells are part of a group known as CD4 cells. Within CD4 cells is another vital sub-group known as regulatory T cells (formerly called suppressor cells). These modulate the T cell response minimising (usually) the chances of auto-immune responses.

4. APC - Antigen presenting cells. Phagocytic, dendritic or B cells may well act as APC's. Thus,

    (i) Macrophages may well destroy an invading microbe but then hold alien antigen in a region of their outer membrane called the **major histocompatibility complex** (MHC). The antigen is then displayed to T cells enabling them to develop an effective response at an early stage of the infection. The macrophages also stimulate T helper cells (chiefly Th2) to secrete chemicals called interleukins, which promote B cell proliferation.

(ii) Dendritic cells present antigen to T **and** B lymphocytes

(iii) B lymphocytes – are also capable of processing and presenting antigens to T lymphocytes.

5. Primary and secondary immune responses. It takes a little while for a response to a new pathogen to be produced. Thus the primary response is weaker than subsequent responses, where Th cells and memory cells will respond immediately. The diagram summarises the differences.

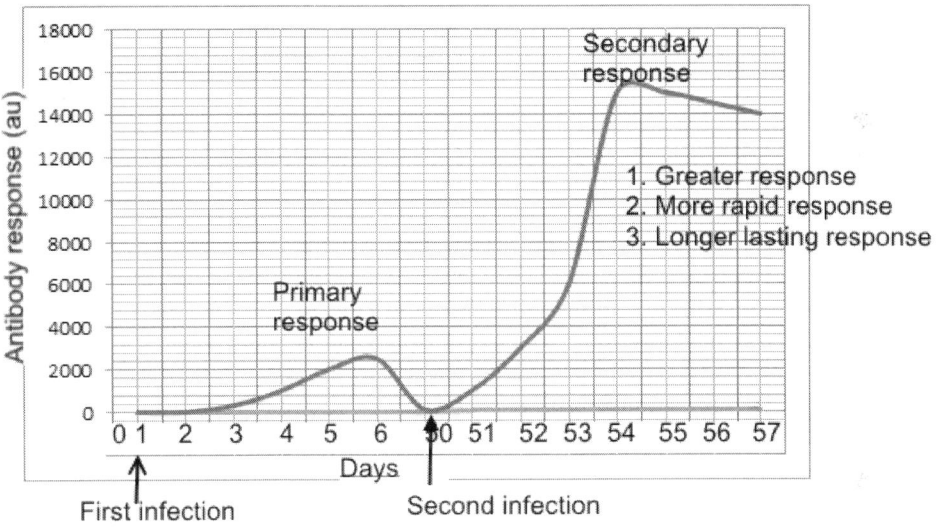

6. Active and passive immunity – comparison of the terminology

	Active immunity	Passive immunity
Naturally acquired	Production of specific antibodies/T killer cells to destroy a pathogen	Receive ready made antibodies. E.g. across placenta, in colostrum
Artificially acquired	Vaccination- being given attenuated or live pathogen OR the antigens (envelope coat vaccine)	Injection of ready made antibodies where there is insufficient safe time for an active response

57

7. The main response to an initial viral infection involves T-lymphocytes. Once T-memory and Th cells have been produced however, the B lymphocyte system will also respond to a repeat viral incursion. Apart from the usual antibody effects, the antibodies bind to viral antigen blocking it from binding with the receptors on our body cells. This is the same principle used by **fusion inhibiting** antiviral drugs (see later).

Medical interventions to help the immune system

As understanding of disease and immunity grows, so does the range of pharmaceutical interventions that can be used.

Antibiotics

Originate from substances produced by micro-organisms to protect them against attack by other micro-organisms. The best known is penicillin which was developed at the beginning of WWII. Modern technology has allowed production of these natural substances to be carried out commercially on a huge scale. Excitingly, other substances with antibiotic qualities are being identified – for instance the new antibiotic **Brilacidin** has been developed based on substances called defensins in the human immune system. But how do antibiotics work?

Most textbooks indicate that antibiotics may act in one of two ways – they may be bactericidal (killing bacteria outright) or bacteriostatic (preventing multiplication of bacteria). However it has become evident that this is a blurred boundary. In some cases some bacteriostatic antibiotic can act in a bactericidal fashion and vice versa.

Different antibiotics work in different ways; some disrupt the bacterial cell wall or membrane, others interfere with bacterial protein synthesis or cell metabolism. Some (e.g. Trimethoprim) prevent the bacterium taking up vital nutrients (e.g. PABA).

The following diagram gives some examples.

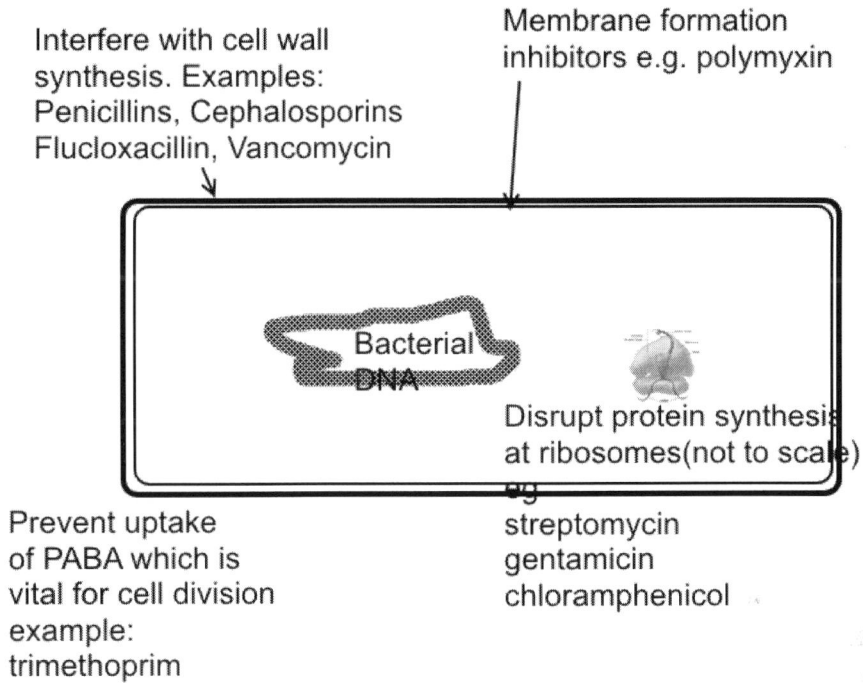

Interfere with cell wall synthesis. Examples: Penicillins, Cephalosporins Flucloxacillin, Vancomycin

Membrane formation inhibitors e.g. polymyxin

Bacterial DNA

Disrupt protein synthesis at ribosomes(not to scale)

Prevent uptake of PABA which is vital for cell division example: trimethoprim

streptomycin gentamicin chloramphenicol

Anti virals

Antibiotics work by disrupting cell function, which is effective for bacteria. However viruses are not cellular and therefore antibiotics are not appropriate.

The first antiviral medications were developed in the 1950's. The arrival of HIV was a huge spur to drug development and study of the various viral infection cycles (described earlier) has enabled drugs to be developed to disrupt the normal cycle.

Some of the possibilities are shown in the following generalised diagram. Exact medication will depend on whether the virus is DNA or RNA, DS or SS.

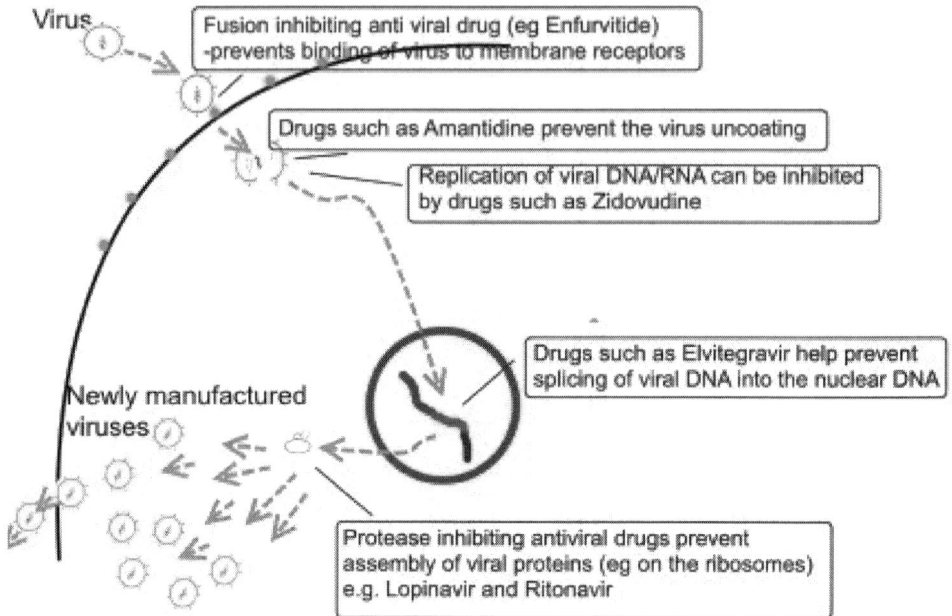

Virus

Fusion inhibiting anti viral drug (eg Enfurvitide) -prevents binding of virus to membrane receptors

Drugs such as Amantidine prevent the virus uncoating

Replication of viral DNA/RNA can be inhibited by drugs such as Zidovudine

Drugs such as Elvitegravir help prevent splicing of viral DNA into the nuclear DNA

Newly manufactured viruses

Protease inhibiting antiviral drugs prevent assembly of viral proteins (eg on the ribosomes) e.g. Lopinavir and Ritonavir

Anti fungals

Fungi are eukaryotes, unlike bacteria (prokaryotic) and viruses (acellular). Interestingly the first identified antibiotic came from the penicillium fungus. Fungal infections may be superficial (i.e. on the body surfaces) or systemic (e.g. spreading into body systems).

Superficial fungal infections are conveniently divided into two classes, candidiasis and dermatomycoses. Examples are thrush and ringworm respectively.

Systemic infections are common – pulmonary aspergillosis affects lungs and some sub-types will spread from the lungs throughout the body (e.g. in immunocompromised subjects).

There are still a lot of unanswered questions about the way the immune system responds to fungal attack. Fungi do have antigens and it is known that phagocytes respond aggressively to them. It is also known that T-lymphocytes also play a significant role. As noted, fungal infections are of particular concern to those who with a suppressed immune response.

The diagram shows just some of the modes of action of various anti-fungals.

Generalised fungal cell showing modes of action of some commonly used antifungals

The anti-fungal Cancidas interferes with cell wall structure

Membrane integrity is destroyed by some antifungals e.g. Amphotericin B

Mitosis is inhibited e.g. by Griseofulvin

Some drugs e.g. Ancotil inhibit DNA synthesis in fungi

Protein synthesis inhibitors e.g. the Sordarins

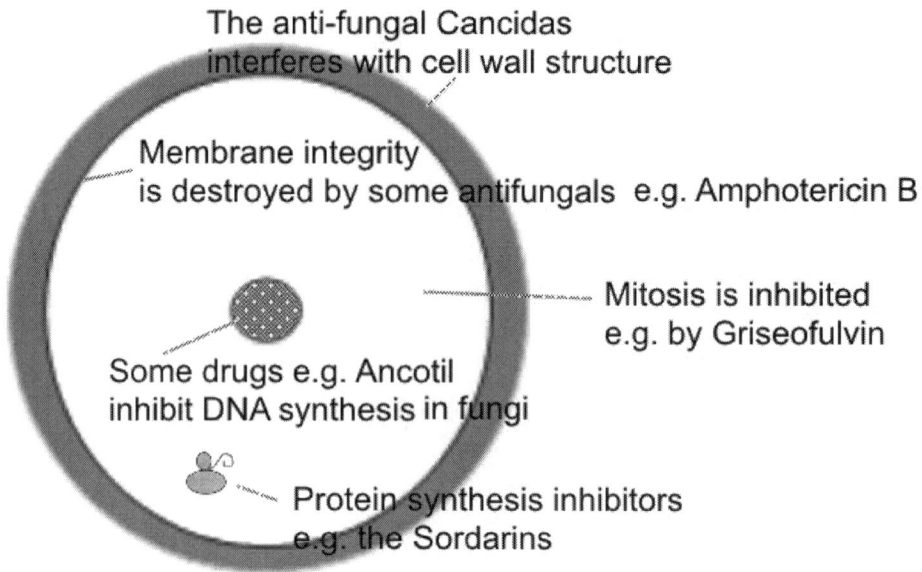

A new approach – quorum sensing and anti-pathogenic drugs

The repeating cycle of antibiotic development→ use→ the appearance of resistant strains is well known. It has led to reluctance on the part of many drug manufacturers to devote millions on R and D for a product which might only have a useful life of a few months before resistance appears.

In the past few years a phenomenon called quorum sensing has been discovered. It has been noted that bacterial populations enter a host and just tick over until they reach a critical mass. The bacteria then release signalling molecules causing an explosion of numbers. Recent research has sought to identify ways of jamming these signals. Promising substances have been noted among the phytochemicals (plant substances). The hope is that the new anti-pathogenic drugs tackling quorum sensing will sidestep the problem of resistance.

Acknowledgements

Diagram of blood cell types: *Blausen.com staff.* "Blausen gallery 2014". *Wikiversity Journal of Medicine.* DOI:10.15347/wjm/2014.010. ISSN 20018762 *[Accessed July 3 2016]*

Diagram of antibody molecule. https://commons.wikimedia.org/wiki/File%3AAntibody.svg [accessed July 4 2016]

Further reading in the series

Basic Introductions to Biology (1) - Cells Tissues and Circulation (also now available in paperback)

Basic Introductions to Biology (2) – Digestion, blood sugar and diabetes

Basic Introductions to Biology (3) - Kidneys, structure and functions

Note that Books 2 and 3 are now also available in a single updated paperback format titled Basic Introductions to Biology. Digestion, blood sugar regulation and the role of the kidneys

Quickfire Revision Questions for Biology Book 1 (e-book or paperback)

Quickfire Revision Questions for Biology Book 2 Going deeper.(e-book or paperback)

A Basic Introduction to Chemistry

Advanced Reading

Lydyard P, Whelan A, Fanger M (2011) *Immunology* Taylor and Francis Abingdon.

Greenstein,Ben (2008) *Trounce's Clinical Pharmacology for Nurses* Elsevier London

References

Bundegard-Nielsen, K. & Nielsen, P. V., 1996. Fungicidal Effect of 15 Disinfectants against 25 Fungal Contaminants Commonly Found in Bread and Cheese Manufacturing. Journal of Food Protection, March, Issue 3, pp. 268-275.

DEFRA, 2011. Review of fungi in drinking water, London: HMSO.

Dix , N. J. & Webster, J., 1995. Fungal Ecology. 1 ed. New York: Springer.

Elsgaard, L. & Prieur, D., 2011. Hydrothermal vents in Lake Tanganyika harbor spore-forming thermophiles with extremely rapid growth. Journal of Great Lakes Research, March, 37(1), pp. 203-206.

Flanagan, P. W., 1978. Microbial ecology and decomposition in arctic tundra and subarctic taiga ecosystems. Microbial Ecology, pp. 161-168.

Gelvin, S. B., 2003. Agrobacterium-Mediated Plant Transformation: the Biology behind the "Gene-Jockeying" Tool. Microbiology and Molecular Biology Reviews, March, Volume 67(1), pp. 16-37.

Londi, P., 2010. Archaeal Ribosomes. [Online]

Available at: http://www.els.net/WileyCDA/ElsArticle/refId-a0000293.html   [Accessed 11 August 2017].

Margulis, L. & Sagan, D., 1986. MICROCOSMOS Four Billion Years of Evolution From Our Microbial Ancestors. Philippines: Summit Books.

☐

Mauriello, E. M. F., Mignot, T., Yang, Z. & Zusman, D. R., 2010. Gliding Motility Revisited: How Do the Myxobacteria Move without Flagella?. Microbiology and Molecular Biology Reviews, June, Volume 74(2), pp. 229-249.

McKay, L. F., Eastwood, M. A. & Brydon, W. G., 1985. Methane excretion in man – a study of breath, flatus, and faeces.. Gut 26, 10.1136/gut.26.1.69 , 26 January.pp. 69-74.

Moore, Moore, D., Robson, G. D. & Trinci, A. P., 2011. 21st Century Guidebook to Fungi. Manchester: Cambridge University Press.

Roossink, M. J., 2015. Move over bacteria! Viruses make their mark as mutualistic microbial symbionts. Journal of Virology.

Warren, J. R. & Marshall, B., 1983. UNIDENTIFIED CURVED BACILLI ON GASTRIC EPITHELIUM IN ACTIVE CHRONIC GASTRITIS. The Lancet, 321(8336), pp. 1273-1275.

Weinberger, L. & Gophna, U., 2015. Archaea in and on the Human Body: Health Implications and Future Directions. PLoS Pathog 11(6): e1004833. doi:10.1371/journal.ppat.1004833, 11 June.

Weissner, D., 2010. The Origins of Viruses. [Online]

Available at: http://www.nature.com/scitable/topicpage/the-origins-of-viruses-14398218   [Accessed July 2017].

Zimmer, C., 2013. An Infinity of Viruses. National Geographic, 20 02.

Glossary

Many of the terms are described in the text; the following expands on the textual information where appropriate.

Allele(s) – the different forms that a single gene can occur in. for example the single gene controlling eye colour in fruit flies can occur in the ebony coloured form (allele) or the red eyed allele.

Anecdotal – evidence is based on personal accounts or experiences rather than hard scientific research. This does not mean that anecdotal evidence is untrue!

Asexual – reproduction involving a single individual. Thus there is no mixing of genetic material with that from another individual. Consequently a line of clones is produced, all genetically identical to the original parent cell. This of course can be complicated if a mutation arises in one of the cell lines. In the case of fungi, asexual means spores.

> Autotrophs – the word means "self-feeder". Autotrophs don't need to find food and break it down. They fabricate the chemicals for cell respiration themselves. Green plants and some bacteria use solar energy to drive photosynthesis, which enables them to make carbohydrates, proteins and lipids from carbon dioxide. They are called photoautotrophs.

Bioremediation – using microorganisms with the ability to break down specific pollutants naturally.

Capsule – a gelatinous outer covering found in some species. There are several chemically different types of capsule. Functions may be defence against predators or to aid adhesion.

Compartmentation- in eukaryotic cells many specialised cell functions are regionalised within membrane bounded organelles.

Conjugation – the process whereby bacteria exchange DNA via pili.

Cytoplasm- of a cell defines everything between the nucleus and the cell membrane. It does not just refer to the fluid component which is called the cytosol.

Diploid   Cells (i.e. gametes) with the usual chromosome number (i.e. in humans two sets of 23 = 46 chromosomes).

Haploid   Cells (i.e. gametes) with HALF the usual chromosome number (i.e. in humans one set of 23).

Heterotrophs/heterotrophic - Heterotrophs need to obtain food and digest it. There are various methods used by different organisms. Briefly the types can be summarised as:

1. Holozoic- Food is taken in and broken down/digested internally. This can be solid food or in liquid form (e.g. aphids feed on sap). There is a wide range of feeding **strategies** in this group.

Hypertonic – a hypertonic solution has a higher concentration of dissolved substances(solutes) than that it is being compared to.

Isotonic – a solution with an equal concentration of solutes to that it is being compared to.

Ma – millions of years ago.

Multifactorial – diseases are more likely to arise when a number of risk factors are present.

Mutations - Mutation is an unpredictable change in the genetic makeup (i.e. the genes in our DNA) of an individual organism.

2. Mutualist – An association exists between two organisms from which both gain some benefit

Mycoplasmas - genus of tiny bacteria lacking any cell walls. In size 0.2-0.3µm. They are parasitic or saprotrophic.

Organoleptic qualities – these are the qualities of food making it attractive (or not). They are qualities of smell, taste, colour and appearance, touch.

Other autotrophs may use chemicals-sulphur for example. These are called chemoautotrophs. They are typically highly specialised bacteria. Some exist in the layers of the earth's crust well away from the surface. Others (Archaea-extremophiles) are found in hydrothermal plumes spewing high temperature sea water (heated deep within the earth) out of vents deep down in the ocean floor.

3. Parasitic – an organism lives in or on another and obtains nutrition from it

Pathogen - a disease causing micro-organism (or parasite e.g. plasmodium for malaria, Schistosomiasis/bilharzia)

Periodontal- diseases caused by gum infections often deep below the gum line. One effect is that an auto-immune response is mounted by the body, resulting in the destruction of healthy bone and soft tissue.

Phyla – in classification, phylum is the next level down after kingdom. The following table gives an example of how it works in the case of humans.

Classification hierarchy

Kingdom	e.g. animalia
Phylum	e.g. chordata
Class	e.g. mammalia
Order	e.g. primates
Family	e.g. hominidae
Genus	e.g. homo
Species	e.g. homo sapiens

Plasmid – a circular loop of DNA. Note this is distinct from the bacterial chromosome which is the bacterial equivalent of the eukaryote nucleus.

Ribosomes – tiny organelles forming a workbench on which amino acids are linked to form proteins (protein synthesis). They are not surrounded by a membrane.

4. Saprobiontic (saprotrophic) – These secrete enzymes and digest externally, then absorb the broken down nutrients. Examples are fungi or decomposing species of bacteria
Saprotrophic- saprotrophs feed on dead or decaying matter.

Sclerotia – produced during adverse conditions and consisting of a thick rinded mass or sphere of hyphae which remain dormant until optimal conditions return.

Thermophilic – bacteria are bacteria that live in extremely hot conditions. Thermophilic bacteria have enzymes that do not denature at these high temperatures.

Thylakoids – membrane bound organelles within cyanobacteria that hold photosynthetic pigments. They are also found within the chloroplasts of green plants at the earth's surface.

Index

Printed in Poland
by Amazon Fulfillment
Poland Sp. z o.o., Wrocław